I0045330

Space Odyssey

A Young Explorer's Guide to the Universe

Welcome to the Space Odyssey!

Have you ever looked up at the night sky, dotted with twinkling stars, and wondered what mysteries the vast universe holds? From the shimmering rings of Saturn to the deep chasms on Mars, space is a realm of endless wonder and excitement.

In this journey, we'll traverse through the cosmic vastness to uncover the secrets of the universe. We'll visit distant planets, dive into the enigmatic black holes, witness the birth and death of stars, and even speculate about the existence of life beyond Earth.

So put on your astronaut helmets, fasten your seatbelts, and prepare for a thrilling odyssey into the heart of space. Let the adventure begin!

Table of Contents:

Chapter 1
The Wonders of Outer Space

The starry night sky is our window to the wonders of outer space

Introduction to Space

Welcome, young adventurers, to the most amazing journey you'll ever take—through the vast and mysterious cosmos!

Imagine a place where there's no air to breathe and no gravity to hold you down. Space is a vacuum, a giant emptiness filled with stars, planets, and cosmic wonders. It's a place where astronauts float weightlessly and where our Earth is just a tiny blue dot among countless other celestial bodies.

As we begin our adventure, let's look up at the night sky. Do you see those twinkling dots of light? Those are stars, just like our Sun but much, much farther away. Some are so big that they could swallow up our whole solar system! And there are planets out there, too, like Jupiter and Saturn, with their amazing rings. We'll also learn about the dazzling constellations that light up our night sky.

Can you spot Orion the Hunter or the Big Dipper? These patterns of stars have stories from ancient times. And don't forget the mysteries of distant planets like Mars, which scientists are exploring to see if life ever existed there. So, get ready for an incredible journey into outer space, where adventure and mystery await us at every turn!

Stars light up the night sky with their sparkling brilliance.

Stars in the Night Sky

Have you ever looked up at the night sky and wondered about those sparkling lights? Well, they're not just lights—they're stars!

Stars are like our Sun, but they're much, much farther away. Some are so huge that they could fit millions of Earths inside them! Stars come in all colors and sizes. Some are red, some are blue, and others are yellow like our Sun.

They twinkle because their light has to pass through Earth's atmosphere before it reaches us. Did you know that stars have their own life stories?
They're born from clouds of gas and dust, and they can shine for millions or even billions of years. But eventually, they run out of fuel and might explode in a huge burst of light called a supernova. Some stars are so heavy that they collapse in on themselves and become black holes.

These are like cosmic vacuum cleaners that suck up everything around them, even light! Stars are like the night's jewels, and they've been guiding explorers and storytellers for centuries. So, the next time you look up at the starry sky, remember that each twinkle has its own amazing story to tell.

Connecting the dots: The art of recognizing constellations and exploring the planets that visit our night sky.

Constellations and Planets

Have you ever tried to connect the dots in the night sky? That's exactly what ancient astronomers did, and they created pictures in the stars called constellations. These aren't real pictures, but they're fun to imagine.

One of the most famous constellations is Orion the Hunter. See those three stars in a row? That's his belt. And he's holding a club and a shield. People from different cultures made up stories about these constellations, and they passed these stories down through generations.

But stars aren't the only things you can see in the night sky. There are also planets! Planets are like big round rocks that go around the Sun, just like our Earth. Some of them are really bright, and you can see them with your own eyes. Venus, for example, is often called the Evening Star because it's so shiny.

There are more planets in our solar system, like Jupiter, Saturn, and Mars, and they all have their own unique qualities. Jupiter has a big red spot that's actually a giant storm, and Saturn has beautiful rings made of ice and rock. So, when you're stargazing, look for the constellations and the planets. Who knows what stories and adventures you'll discover in the night sky!

Chapter 2
Mars: The Red Planet

Journey to Mars: Challenges, adventures, and missions to the Red Planet.

Journey to Mars

Hey there, future astronauts! Buckle up because we're embarking on an epic journey to Mars, the Red Planet! Why Mars, you ask? Well, it's the most Earth-like planet in our solar system.

But reaching Mars isn't a walk in the park; it's a thrilling adventure! Imagine soaring through space in a rocket, leaving Earth's atmosphere behind. You'll see the stars twinkle, and the Earth will become a tiny blue dot in the rearview mirror.

As you get closer to Mars, you'll marvel at the rusty-red planet growing larger in the window. Landing on Mars is a whole different challenge. It has a rocky, dusty terrain, and sometimes, you'll need a parachute to slow down. But once you're there, it's like being on another world.

You'll experience weaker gravity, so you can bounce around like an astronaut superhero! Get ready for this incredible journey filled with excitement and discovery!

Martian terrain and weather: A rocky world with towering volcanoes and swirling dust storms.

Martian Terrain and Weather

Welcome to Mars, young explorers! The Martian landscape is like nothing on Earth. Picture yourself on a desert made of rusty rocks and mountains.

It's a bit like the Grand Canyon but redder! One of Mars' cool features is Olympus Mons, the tallest volcano in the solar system. It's so high that it pokes into space! But be ready for some wild weather.

Mars can be a dusty and windy place, with huge dust storms that can cover the whole planet. Imagine standing in a red storm, where the sky turns dark, and the winds howl.

Exploring Mars means understanding its unique terrain and weather. So, grab your spacesuits, and let's uncover the mysteries of the Red Planet!

The quest for life on Mars: Uncovering the mysteries of the Red Planet's past

The Search for Signs of Life on Mars

Could Mars hold the secret to life beyond Earth? That's a question scientists are eager to answer! Mars might have been home to tiny living creatures long ago, or it might still have hidden pockets of life today.

To solve this mystery, we're sending rovers and scientists to explore. Imagine driving a rover on the Martian surface, carefully examining rocks and soil, searching for clues.

And sometimes, we even send robots to dig below the surface to find water ice! Water is a key ingredient for life, and if we find it, it could be a game-changer.

So, get ready to join the quest to uncover signs of life on the Red Planet. Who knows what amazing discoveries await us!

Chapter 3
Black Holes: The Cosmic Mystery

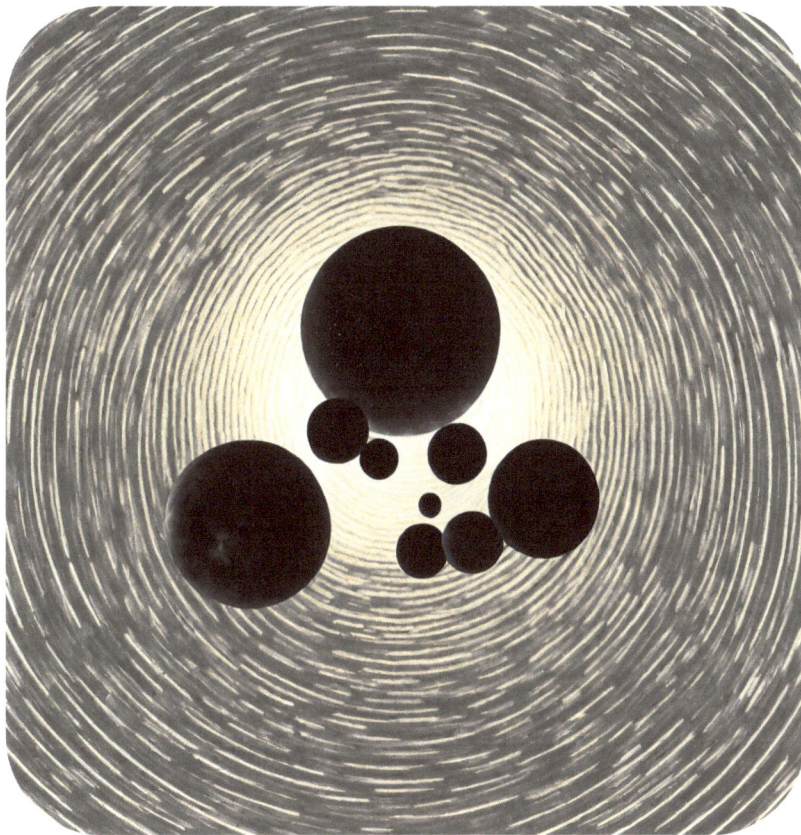

What Are Black Holes? The cosmic vacuum cleaners that devour everything in their path.

What Are Black Holes?

Get ready to journey into the unknown, young space explorers! Black holes are like nothing else in the universe.

Imagine a place where gravity is so strong that nothing, not even light, can escape. That's a black hole! They're like cosmic vacuum cleaners, sucking up everything around them.

But how do they form? Well, when a giant star runs out of fuel, it collapses in on itself. If it's big enough, it becomes a black hole. Black holes come in different sizes, from small to supermassive.

The supermassive ones live at the centers of galaxies and are millions or even billions of times heavier than our Sun!

Now, here's the wild part: if you ever got too close to a black hole, time would slow down for you compared to the rest of the universe.

You could watch as stars and galaxies zoom by while your clock ticks slower. It's like time travel, but with a twist!

The strange effects of black holes: Time warps, space bends, and the universe behaves oddly.

The Strange Effects of Black Holes

Black holes aren't just cosmic vacuums; they create some of the weirdest effects in the universe.

Imagine standing near a black hole, and it's like your watch goes crazy. Time slows down, and you could even see events from the past and future all mixed up!

Also, black holes warp space, making it bend and twist.

This means that if you get too close to one, you might get stretched like spaghetti – that's what scientists call 'spaghettification.'

Don't worry; it's not something you'll experience on your space adventures!

Theories and discoveries about black holes: From the heart of galaxies to the remnants of massive stars.

Theories and Discoveries

Scientists are like space detectives trying to solve the mystery of black holes. They've found some incredible clues!

Supermassive black holes are hiding at the hearts of galaxies, including our Milky Way. And there are stellar-mass black holes, too, formed from the remnants of massive stars.

We're still figuring out how they form and what happens inside them.

Some think they might be doorways to other parts of the universe!

Studying black holes has opened up a whole new world of discoveries, from finding invisible objects to exploring the weirdness of space and time.

Chapter 4
Stars: Birth, Life, and Death

How stars form and shine: The cosmic dance of birth and brilliance.

How Stars Form and Shine

Buckle up, young astronomers!

We're about to dive into the incredible world of stars—those sparkling lights that fill our night sky.

But have you ever wondered how stars are born and why they shine so brightly? Well, it's like a cosmic dance in space! Imagine vast clouds of gas and dust floating through the galaxy.

These clouds start to squeeze together under gravity's embrace, creating a pressure cooker where temperatures rise.

When it gets hot enough, something magical happens—nuclear fusion!

It's like a never-ending party where hydrogen atoms fuse together to make helium, releasing a burst of energy in the process.

That energy is what makes stars shine! And just like that, a new star is born. But this is just the beginning of their incredible journey through space.

Stellar evolution and supernovae: The epic tale of stars' transformation and explosive finales.

Stellar Evolution and Supernovae

Stars are like cosmic superheroes, and they have their own life stories!

They go through stages just like people. As they grow older, they change and evolve. Some stars, called supergiants, swell up to hundreds of times their original size!

And then comes the grand finale—the supernova. It's like a star's explosion party!

The explosion is so powerful that for a brief moment, a single star can outshine an entire galaxy. This explosion spreads elements like carbon, oxygen, and even gold into space.

These elements are the building blocks of planets and, yes, even us! The life cycle of a star is a cosmic drama, and we're here to unravel its secrets.

The life cycle of stars: A cosmic rhythm that shapes the universe.

The Life Cycle of Stars

Stars are like cosmic timekeepers, and their lives have a rhythm. It all starts when a star is born, and it continues through various stages.

As stars get older, they change colors, sizes, and even personalities! Some become red giants, while others become white dwarfs.

And then there are the superstars—massive stars that end their lives in spectacular explosions called supernovae. After the fireworks, they leave behind remnants like neutron stars or even black holes.

The story of stars is a grand cosmic journey that impacts the entire universe, shaping galaxies and seeding space with the stuff of life.

So, young stargazers, get ready to witness the epic life cycle of stars, from birth to grand finales!

Chapter 5
The Solar System

Our Sun: The ultimate star that powers our solar system.

Our Sun: The Ultimate Star

Welcome to the heart of our solar system, where all the action begins—the Sun!

Our Sun isn't just any star; it's the star that lights up our entire world.

Think of it as a gigantic, glowing, and super-hot ball of gas. It's so massive that it holds the whole solar system together with its powerful gravity.

And the Sun isn't just a giant lightbulb; it's a dynamic powerhouse. It's like a cosmic campfire, where hydrogen atoms are constantly turning into helium and releasing tons of energy in the process.

This energy is what warms our planet and gives us daylight.

Without the Sun, there would be no life on Earth! So, let's get up close and personal with the Sun, the ultimate star of our show.

The inner planets: Rocky worlds that orbit close to the Sun.

The Inner Planets: Mercury, Venus, and Earth

Now that we've met the Sun, let's journey to the planets that orbit closest to it.

These are the inner planets—Mercury, Venus, and Earth. They're like the Sun's best buddies, and they're rocky and solid just like our home planet.

First up, Mercury, the speed racer of the solar system. It's super hot during the day and freezing cold at night!

Next, Venus, the hottest planet with a runaway greenhouse effect. And finally, our beloved Earth, the perfect oasis for life as we know it.

It's a blue-and-green wonderland filled with oceans, mountains, and living creatures.

Together, these inner planets have their own unique personalities, and Earth is the crown jewel. Let's explore the rocky neighbors of the Sun!

The outer planets: Giants and protectors of our solar system's cosmic balance.

The Outer Planets: Jupiter, Saturn, Uranus, and Neptune

But wait, there's more to our solar system! Beyond the inner planets, we have the outer planets—Jupiter, Saturn, Uranus, and Neptune.

These giants are like the solar system's protectors, and they're very different from the inner planets. Imagine a planet so massive that it could fit all the other planets inside it—that's Jupiter! It's a stormy giant with a colossal red spot.

Saturn is famous for its stunning rings made of ice and rock. Uranus and Neptune are the mysterious twins with icy blue atmospheres.

These outer planets are full of surprises, from raging storms to icy moons. They're like the superheroes of the solar system, and they keep the cosmic balance.

Get ready to journey to the outer reaches of our solar neighborhood!

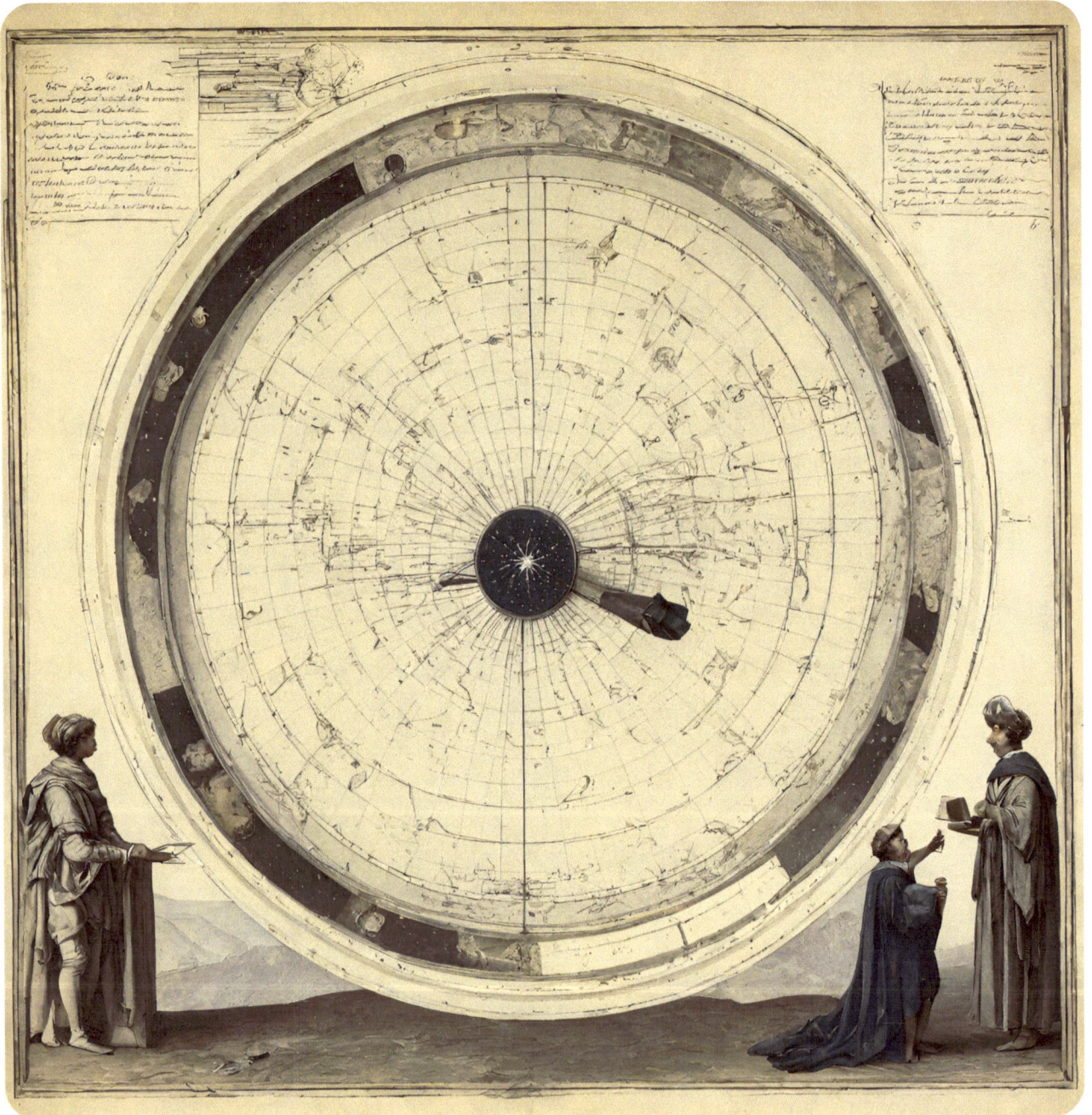

Chapter 6
Astronomy Through History

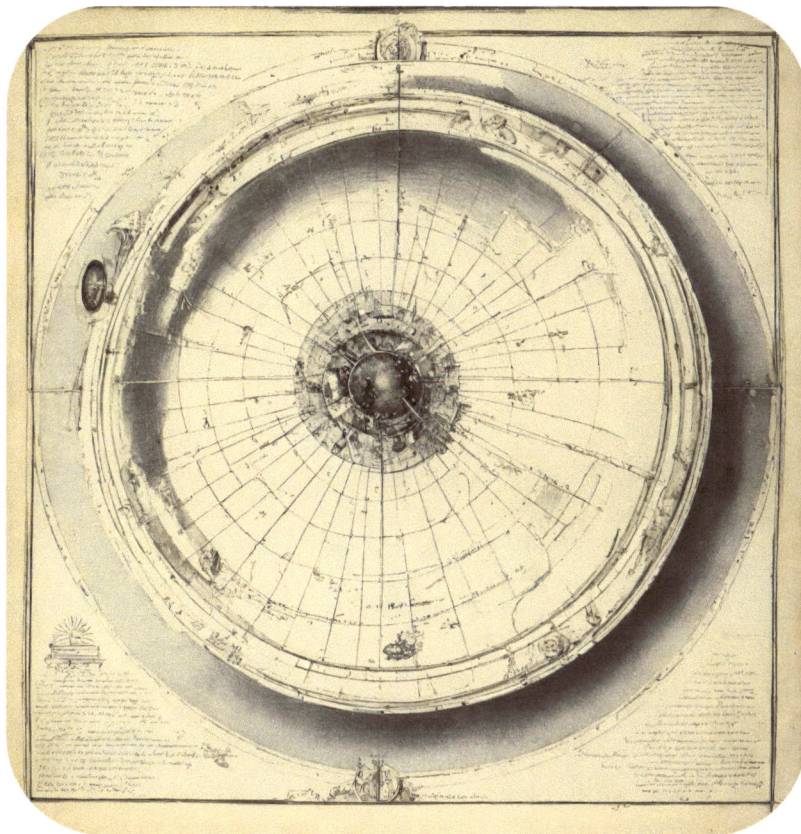

Early observations of the night sky:
Discovering the magic of the cosmos
through ancient eyes.

Early Observations of the Night Sky

Travel back in time, young stargazers, to when our ancestors first looked up at the night sky in wonder.

Long before telescopes and rockets, people were astronomers too!

They told stories about the stars, connected the dots to make constellations, and used the Moon to mark the passage of time.

Can you imagine navigating using the stars like ancient mariners? They believed the stars had magical powers and stories to tell.

These early observations were the beginnings of our cosmic journey, and they still inspire us today.

So, let's look up at the same stars our ancestors did and discover the magic of the night sky!

Famous astronomers and their discoveries: Pioneers of cosmic exploration.

Famous Astronomers and Their Discoveries

Fast forward to the days when telescopes changed everything.

Meet the cosmic detectives—the famous astronomers who unlocked the secrets of the universe.

Galileo Galilei, the Italian scientist, used a telescope to peer at the Moon and Jupiter, discovering moons orbiting another planet!

Johannes Kepler cracked the code of planetary motion, and Isaac Newton showed us how gravity keeps us grounded.

Caroline Herschel was one of the first female astronomers who discovered comets, while Edwin Hubble found that the universe is expanding.

These astronomers and many others paved the way for modern space exploration.

Their discoveries are like pieces of a cosmic puzzle that we're still putting together. Get ready to be inspired by their tales of stargazing and discovery!

Modern space exploration: The exciting journey to understand the cosmos.

Modern Space Exploration

We're living in the most exciting era of space exploration!

Humans have built incredible telescopes, sent robots to other planets, and even walked on the Moon.

Our quest to understand the universe has taken us to places beyond our wildest dreams.

Telescopes like the Hubble Space Telescope have shown us breathtaking images of distant galaxies and nebulae. Rovers like

Curiosity are exploring the surface of Mars, sending back selfies from another planet!

And astronauts on the International Space Station conduct experiments in microgravity, helping us learn about life in space.

We're uncovering cosmic mysteries and taking steps to become a spacefaring species.

The adventure continues, and maybe one day, you'll be the next space explorer, discovering new frontiers and making history!

Chapter 7
Space Travel and Astronauts

The Space Race: From Earth to the Moon and an epic leap for humankind.

The Space Race: From Earth to the Moon

Imagine a time when the world was caught in an epic space race!

It was a competition like no other, with two superpowers, the United States and the Soviet Union, racing to reach the Moon.

It all began with a dream and a challenge—President John F. Kennedy's promise that we'd land a person on the Moon and bring them back safely.

The Apollo missions became our ticket to the Moon, and on July 20, 1969, Apollo 11 made history.

Astronauts Neil Armstrong and Buzz Aldrin stepped onto the lunar surface while Michael Collins orbited above. It was a giant leap for humankind!

The Space Race taught us that with determination and teamwork, we can achieve the impossible. It was just the beginning of our cosmic adventures!

Living and working in space: Astronauts' extraordinary life among the stars.

Living and Working in Space

Now, let's fast forward to today, where astronauts are living and working in space like real-life cosmic pioneers.

The International Space Station (ISS) is their home away from home, orbiting Earth at incredible speeds.

Astronauts float weightlessly, conduct experiments, and even grow plants in microgravity!

Living on the ISS isn't all fun and games; it's a challenging job that requires teamwork, discipline, and training.

Astronauts become scientists, engineers, and even chefs in space. They eat special space food and drink water droplets that float in the air!

Living in space isn't just about the Moon and Mars; it's about understanding how humans can survive in the cosmos.

So, young space enthusiasts, get ready to learn what it's like to live and work among the stars!

The future of space travel: Exploring new frontiers and reaching for the stars.

The Future of Space Travel

What's next for space travel?

The future is filled with exciting possibilities! We're planning missions to the Moon, Mars, and even beyond our solar system.

NASA's Artemis program aims to return humans to the Moon, including the first woman and the next man!

Mars missions are on the horizon, where astronauts will step foot on the Red Planet.

We're also dreaming of futuristic spacecraft like space shuttles and powerful rockets that can take us to distant stars.

The cosmos is our playground, and the adventure never ends.

So, young space dreamers, get ready to explore the bright future of space travel, where the sky is not the limit—it's just the beginning!

Chapter 8
Extraterrestrial Life and the Search for Aliens

Are we alone in the universe? The cosmic mystery waiting to be solved.

Are We Alone in the Universe?

Hey there, young cosmic detectives!

One of the biggest questions in science is: Are we alone in the universe?

The answer? We don't know...yet!

The universe is mind-bogglingly vast, with billions of galaxies, each containing billions of stars.

It's like having a gigantic cosmic playground! With so many stars, there's a chance that other planets, like Earth, exist out there. And where there are planets, there could be life.

Scientists are on a quest to find evidence of extraterrestrial life, whether it's tiny microbes on Mars or intelligent aliens in distant galaxies.

The universe is a treasure trove of mysteries waiting to be uncovered. So, get ready to explore the possibilities of life beyond our planet!

SETI: The search for extraterrestrial intelligence and the cosmic radio detectives.

SETI: The Search for Extraterrestrial Intelligence

Imagine being a cosmic radio detective, searching the skies for messages from other intelligent beings.

That's what scientists at SETI (Search for Extraterrestrial Intelligence) do!

They use huge radio dishes to listen for signals that might come from distant civilizations.

It's like tuning in to an interstellar radio station. They've been scanning the cosmos for decades, but so far, they haven't received a call from ET.

But the search continues because the universe is so vast that we've only scratched the surface.

Maybe one day, we'll hear a cosmic hello from another world. Until then, we're like space detectives, always on the lookout for signs of alien neighbors.

Get ready to join the search for extraterrestrial intelligence!

Exoplanets: Earth-like worlds beyond our solar system and the quest for alien cousins.

Exoplanets: Earth-Like Worlds Beyond Our Solar System

Hold on to your space helmets because we're about to discover exoplanets—planets that orbit stars beyond our solar system!

Scientists have found thousands of these alien worlds, and some of them might be a lot like Earth. They're not too hot, not too cold, and just right for life as we know it.

These exoplanets are like distant cousins of Earth, and they're scattered throughout the galaxy.

Imagine a journey to one of these exoplanets where you'd see unique landscapes and skies. Some might have two suns! Scientists are on the lookout for planets that could host life, and they're finding some pretty exciting candidates.

The universe is full of surprises, and exoplanets are the next frontier in our search for alien neighbors.

So, let's blast off and explore these Earth-like worlds beyond our solar system!

Chapter 9
Space Technology and Telescopes

How telescopes work: Unlocking the secrets of the cosmos with cosmic eyes.

How Telescopes Work

Ever wondered how we peek into the depths of space?

It's all thanks to the incredible invention called the telescope!

Telescopes are like cosmic eyes that help us see stars, planets, and galaxies that are millions of light-years away. But how do they work their magic? It's all about gathering light.

Telescopes have big mirrors or lenses that collect and focus light from distant objects. This light then gets magnified, making things appear closer and clearer.

Imagine having super-vision that lets you see things you couldn't see with your naked eye!

Telescopes come in all sizes, from backyard ones to colossal space telescopes like the Hubble.

They've revealed wonders beyond our imagination, from distant galaxies to the rings of Saturn.

So, young stargazers, get ready to discover the secrets of how telescopes unlock the mysteries of the universe!

Space observatories and their discoveries: Unveiling the secrets of the universe.

Space Observatories and Their Discoveries

Imagine having a superpower to see things even when they're too far away for regular telescopes.

That's what space observatories do! These are telescopes and scientific instruments placed in space, far above Earth's atmosphere.

Why space? Because our atmosphere blurs and blocks some of the light from space. Space observatories can see clearer and farther.

They've made groundbreaking discoveries about the universe. One of the cosmic heroes is the Hubble Space Telescope. It's like the world's most powerful camera in space.

Hubble has captured stunning images of galaxies, nebulae, and planets. It's shown us the birth of stars and the edge of the observable universe!

And there are more space observatories out there, each with a unique mission. They're like cosmic detectives, helping us understand the mysteries of the cosmos.

So, let's embark on a cosmic journey to explore these space observatories and the wonders they've unveiled!

The Hubble Space Telescope: Unveiling the universe and rewriting the history of the cosmos.

The Hubble Space Telescope: Unveiling the Universe

Prepare to be awestruck by one of the greatest space heroes—the Hubble Space Telescope!

It's not just a telescope; it's a time machine that has transported us to the farthest reaches of space and time. Hubble orbits high above Earth, capturing breathtaking images of the cosmos.

It's shown us galaxies colliding, the birth of stars, and the beauty of nebulae. But that's not all; it's also measured the expansion rate of the universe, helping us understand the big bang!

Hubble has gazed into the past, letting us see galaxies as they were billions of years ago. It's like a cosmic storyteller, revealing the history of the universe.

The Hubble Space Telescope has forever changed our view of the cosmos.

So, young space explorers, get ready to dive into the mesmerizing world of Hubble and its mind-blowing discoveries!

Chapter 10
The Mysteries of Dark Matter and Dark Energy

What is dark matter? The invisible cosmic enigma that defies our senses.

What Is Dark Matter?

Imagine a cosmic puzzle, young space adventurers!

Scientists have discovered that most of the universe is made of something mysterious and invisible called 'dark matter.'

It's like an invisible cloak that wraps around galaxies, making them spin faster than they should. But here's the twist—dark matter doesn't emit light or energy, so we can't see it directly!

It's a cosmic enigma that challenges our understanding of the universe. Scientists are on a quest to uncover the secrets of dark matter, and they use powerful telescopes and detectors to search for its clues.

Dark matter is like the hidden treasure of the cosmos, waiting to be found!

The enigma of dark energy: The cosmic force that's tearing the universe apart.

The Enigma of Dark Energy

Hold on tight, because the mysteries of the cosmos are getting even stranger!

Dark energy is another cosmic riddle that's pushing the universe apart faster and faster. It's like an anti-gravity force that's making galaxies drift apart at an astonishing rate.

What's even more mind-boggling is that dark energy makes up most of the universe's energy content. But, just like dark matter, it's invisible and elusive.

Scientists are racing to understand dark energy, and they're using powerful space telescopes and experiments to study its effects on the universe.

Dark energy is like the cosmic wind blowing galaxies apart, and it's a force we're only beginning to grasp.

Get ready to explore the enigma of dark energy, where the universe's fate hangs in the balance!

The role of dark matter and dark energy in the universe: Cosmic enigmas that shape our cosmos.

The Role of Dark Matter and Dark Energy in the Universe

Dark matter and dark energy might sound like villains from a space adventure, but they're essential players in the cosmic drama of the universe!

Dark matter acts as cosmic glue, holding galaxies and galaxy clusters together with its invisible gravitational force. Without it, the universe would look very different.

On the other hand, dark energy is the mysterious force responsible for the universe's ever-accelerating expansion. It's like the cosmic accelerator pedal!

Together, these cosmic enigmas shape the fate of the universe. Scientists are like cosmic detectives, trying to solve the puzzle of dark matter and dark energy to understand the past, present, and future of our cosmos.

So, young space explorers, get ready to dive into the thrilling mysteries of these invisible forces that govern the universe!

Glossary

1. **Astronomer:** A scientist who studies stars, planets, galaxies, and other objects in space.
2. **Constellation:** A group of stars that forms a pattern or shape in the night sky, like the Big Dipper or Orion.
3. **Galaxy:** A vast collection of stars, planets, gas, and dust held together by gravity, like our Milky Way.
4. **Microgravity:** A condition in which objects seem to float because they are experiencing very weak gravity, like astronauts in space.
5. **Orbit:** The path that an object, like a planet or satellite, follows as it goes around another object in space, like the Earth orbits the Sun.
6. **Solar System:** Our Sun and all the objects that orbit it, including planets like Earth and Mars.
7. **Telescope:** A device that helps us see distant objects in space by gathering and magnifying light.
8. **Spacecraft:** A vehicle designed to travel in outer space, like rockets, shuttles, and probes.
9. **Asteroid:** A small rocky object that orbits the Sun, often found in the asteroid belt between Mars and Jupiter.
10. **Meteoroid:** A small rock or metal object in space, smaller than an asteroid, that can become a meteor when it enters Earth's atmosphere.
11. **Black Hole:** A mysterious region in space with gravity so strong that nothing, not even light, can escape from it.
12. **Extraterrestrial:** Anything that originates from outside Earth, like aliens or objects from space.
13. **Gravity:** The force that pulls objects toward each other, like how Earth's gravity keeps us on the ground.
14. **Planet:** A large object that orbits a star, shines by reflecting its star's light, and can have its own moons.
15. **Star:** A massive, glowing ball of gas that produces heat and light through nuclear reactions, like the Sun.
16. **Universe:** Everything that exists, including all the galaxies, stars, planets, and space.

A Special Note

Hello Fantastic Space Explorers! 🚀

I hope you had an incredible time journeying through the cosmos with us! Space is a universe filled with endless marvels, and you're now a certified space adventurer.

Exploring the mysteries of outer space has been an absolute delight. Just like we love unveiling the wonders of the world in the most exciting way, we've loved sharing the magic of the universe with you.

If you can spare a moment, we'd be over the moon if you could share your thoughts on Amazon or wherever you picked up this book. Your feedback isn't just a guiding star for fellow space explorers; it also ignites our passion for creating more adventures that ignite your imagination.

Thank you for joining us on this epic journey through the cosmos. Keep gazing at the stars, keep dreaming big, and always treasure your enchanting voyages among the galaxies!

With boundless curiosity and best wishes,

Dr. Amit Khanna
Founder, Biotic Chronicles

About the Author

Dr. Amit Khanna is a distinguished molecular biologist and a storyteller at heart. With a rich background in genomics, he navigates the intricate dance of molecules, genes, and DNA. But for Amit, it's more than just the science; it's the stories that lie hidden within our genetic code, the tales that nature has silently penned down over eons.

While many are content with the black and white of scientific data, Dr. Khanna sees a vivid tapestry of narratives. Beyond his research, he has a unique talent for narrating science in a way that captivates, educates, and inspires. He brings the wonders of the molecular world to life, making it accessible and fascinating for audiences of all ages.

Whether it's through his books or his engaging science narrations, Dr. Khanna's mission is to bridge the gap between complex biological phenomena and the curious minds eager to understand them.

To learn more about his narrations, explorations, and the world of biology through his lens, visit www.BioticChronicles.com.

Dive Deeper with Biotic Chronicles!

For a delightful collection of books tailored for children aged between 5 and 15 years old, as well as exclusive behind-the-scenes content, downloadable coloring pages, and engaging monthly newsletters, chart your course to www.BioticChronicles.com. Immerse yourself in the magical and scientific realms waiting to be explored!

Important Note About Illustrations:

Dear Space Explorers,
Please note that the illustrations in "Space Odyssey" are creative representations and not scientifically generated images. They aim to make space exploration exciting and imaginative. Enjoy your cosmic journey!

Copyright Notice:

All rights reserved. No part of this book may be reproduced or transmitted in any form or by any means, electronic or mechanical, including photocopying, recording, or any information storage and retrieval system, without prior written permission from the publisher.

www.ingramcontent.com/pod-product-compliance
Lightning Source LLC
Chambersburg PA
CBHW052052190326
41519CB00002BA/189